THE LIBRARY OF
FUTURE ENERGY

HYDROPOWER
OF THE FUTURE
NEW WAYS OF TURNING WATER INTO ENERGY

ALLISON STARK DRAPER

THE ROSEN PUBLISHING GROUP, INC.
NEW YORK

BPMS MEDIA CENTER

Published in 2003 by The Rosen Publishing Group, Inc.
29 East 21st Street, New York, NY 10011

Library of Congress Cataloging-in-Publication Data

Draper, Allison Stark.
Hydropower of the future: new ways of turning water into energy/by Allison Stark Draper.— 1st ed.
      p. cm. — (Library of future energy)
Summary: Discusses the pros and cons of hydroelectric power, including an overview of its history in the United States and other countries' attempts to balance electric needs and the environment.
Includes bibliographical references and index.
ISBN 0-8239-3664-3 (library binding)
1. Hydroelectric power plants—Juvenile literature. 2. Water-power— Juvenile literature. [1. Hydroelectric power plants. 2. Water power.]
I. Title. II. Series.
TK1081 .D68 2002
333.91'4—dc21

                                                          2001008725

Manufactured in the United States of America

CONTENTS

INTRODUCTION

The population of Earth grows larger every day. More people need more and more energy. Coal and oil, our primary energy sources, are fossil fuels. They formed over millions of years from heat and pressure inside Earth. We cannot replace them. Even worse, when we burn coal and oil, we pollute the air and cause global warming.

Alternative energy sources, such as water power, are safer and cleaner. They are sustainable, which means we can use them over and over again. Water power does not use up water. It uses the energy of rushing water to create electricity. Water power is the world's leading source of renewable energy. It produces more than 95 percent of sustainably

Smoke rises from a factory burning coal in Sichuan, China. Emissions from burning fossil fuels—coal, petroleum, and natural gas—are a main cause of air pollution and global warming.

generated electricity. Other sustainable energies include solar, wind, and geothermal power. Water power is the most ancient method, and so far it is the most practical of all. The modern name for water power that generates electricity is hydropower.

People have used water power for thousands of years. The ancient Greeks harnessed the force of rushing rivers to power waterwheels that crushed grapes for wine and ground grain for bread. "Hydro" comes from the Greek word meaning water or liquid. Thirteenth-century Chinese engineers built machines that used the tidal power of waves to crush iron ore. In the fifteenth century,

Italian painter and inventor Leonardo da Vinci designed a wave machine. By the nineteenth century, several coastal nations had plans to fuel their growing industries with sea power.

In the twentieth century, gas and oil fuel became inexpensive and plentiful. Experiments with sustainable energy sources, like the ocean, stopped because they were so costly. Interest in alternative energy rose

Humans have been using water power for centuries. This illustration of a horizontal waterwheel with complex gearing and screws appeared in a book that was written in 1588.

briefly during the oil crisis of the 1970s. But the crisis ended, and people continued to use gas and oil through the eighties and nineties. By the end of the twentieth century, there was a looming shortage of fossil fuels. When the advance of global warming made it clear to some people that we needed to find cleaner fuels, there was renewed interest in the idea of making power from water.

Water spills over a waterfall or rushes through a dam with tremendous power. The distance water drops to the point where humans retrieve its energy—with a waterwheel or a hydropower turbine—is called the head. The longer the drop, the greater the head.

Flow is the amount of water that falls. A wide, deep river provides lots of flow. The more flow, the more force. This is why a person can stand under a stream's waterfall, but would get crushed by the waterfall of a large river. It was a great moment in human ingenuity when someone discovered that the force of water could be put to work.

The simplest, oldest way to convert water to energy is with a mill wheel. In the early days of water power, waterwheels turned

Water from a trough powered this two-story-high waterwheel on a farm in Missouri during the early 1900s.

with enough power to grind grain, saw logs, or pump water.

Similar to a Ferris wheel, a waterwheel is a large circle of wood with blades or cups attached to its rim. Sitting in a stream or at a narrow point below a waterfall, it traps water flowing over the fall in the cups of the wheel. The force of the water presses on the cups and turns the wheel.

A waterwheel turns slowly, connecting to a much smaller wheel. Receiving the power of the large, slow wheel, the smaller wheel turns much faster. Called gearing, this is similar to gears on a bicycle that enable a rider to pedal more powerfully up hills or faster on flat surfaces.

If a river does not contain a point narrow enough for a waterwheel, a trough or a pipe can divert a stream of water and spill it onto the waterwheel. If a river is fast but flat, a pipe can carry the water downhill and force it through a nozzle that shoots a very hard stream of water at a turbine. A turbine, like a small, fast waterwheel, can power an electric generator directly, without gears.

HYDROLOGY AND GRAVITY

The world's water evaporates from lakes and oceans, and condenses into clouds. Gravity pulls it back to the earth as rain or snow. Water runs down from mountains and highlands into lakes, rivers, and oceans. This is called the hydrologic cycle. It is driven by the Sun. Closed and constant, the hydrologic cycle does not lose or gain water. It transforms water from liquid to gas and back to liquid.

THE HYDROPOWER PLANT

Most hydropower plants have three parts. A reservoir stores water above a drop, to create head. A dam opens and closes to control water flow. A power plant transforms the moving energy of water into mechanical energy or electricity.

When the dam opens, water flows from the reservoir through a large tube called a penstock. Turbines are at the bottom of the penstock. The swiftly flowing water presses against the blades inside a turbine and makes it spin.

The spinning turbine turns a shaft connected to the generator. The energy from the spinning shaft turns, or rotates, the rotor. The rotor's electromagnets spin inside the tight copper coil of the stator, a nonmoving part. The motion moves electrons, creating an electric field between the coil and the magnets. Electricity leaves the generator as electric current, traveling through high-voltage power cables to a local utility company that distributes it to homes and offices.

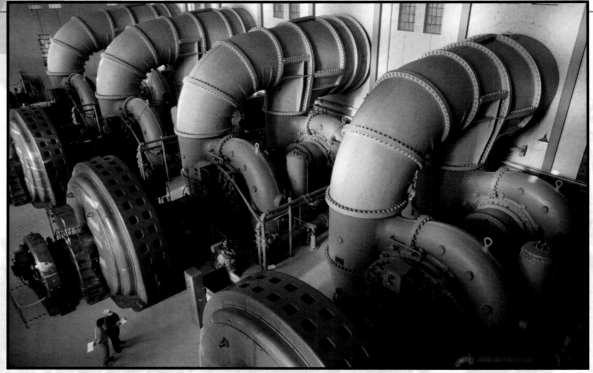
These are turbines at the Avista Corporation's hydropowered Nine Mile Plant along the Spokane River in Washington State. Avista operates eight hydroelectric plants on the Clark Fork and Spokane Rivers that provide electricity for more than 300,000 homes.

There are two basic kinds of turbines that move the machines that make hydropower: impulse turbines and reactive turbines. An impulse turbine moves pressurized steam into a stationary nozzle where it is directed onto blades carried by a rotor. The reactive turbine moves as a result of gas against the blades. The two systems can be combined to create an impulse-reactive turbine.

Natural waterfalls make excellent locations for hydropower plants, but powerful natural waterfalls, like Niagara Falls, are rare. Large rivers are more common. By damming a river, it is possible to create an artificial fall that can be as good or better than a natural fall.

THE HYDROPOWER PLANT

The Alaskan Tazimina Hydroelectric Project is a diversion project. Located 175 miles southwest of Anchorage on Alaska's Tazimina River, it has a production capacity of 1,648 kilowatts.

Engineers can control the head and flow of a dam. If electricity is needed, stored water in the reservoir is available to make more. When heavy rains send too much water down a river, the dam operator can lift gates to release excess water through spillways.

In a pumped storage plant, when the need for electricity drops, the power plant can pump the water back to the upper reservoir. Impoundment hydropower uses a dam to store water in a lake or a reservoir, releasing enough water to meet the demand for electricity. Diversion hydropower does not store water. Instead, it channels river water through a canal or a penstock.

Fossil fuel plants convert 50 percent of raw material to electrical power. In comparison, modern hydropower turbines can turn as much as 90 percent of the energy of its water into electricity. Hydropower costs one-third as much to produce as does power from fossil fuel.

Hydropower is a great benefit for countries rich in waterways, especially if the countries don't have their own reserves of fossil fuel. By using local, renewable resources such as hydropower, a country can avoid world trade crises and fuel shortages.

2 THE HISTORY OF HYDROPOWER

Hydropower dates back more than 2,000 years. The earliest known references to the water mill appear in Greek, Roman, and Chinese texts. They describe vertical waterwheels placed in rivers or streams. As the large wheels turned, they cranked smaller, faster millstones that ground corn into meal.

By the fourth century AD, water milling had begun in other parts of Asia and in northern Europe. In the early eleventh century, William the Conqueror remarked on the thousands of water mills in England. Most of them used stream or river power, but some worked with the tides.

The earliest waterwheels were mounted high so that the water hit the bottom of the

Construction workers examine a horizontal waterwheel generator during assembly at a hydroelectric power station in 1941.

wheel. Later, millers lowered their wheels, diverting streams over their tops. Eventually, engineers began to place the wheels on their sides. This was more efficient because it took less water power to turn the wheel.

In the late 1700s, an American named Oliver Evans designed a mill in which a combination of gears, shafts, and conveyors enabled the wheel to grind grain and carry it around the mill. By the nineteenth century, waterwheels were the power source in sawmills, textile mills, and forges.

In 1826, the French engineer Jean-Victor Poncolet envisioned an even more efficient waterwheel. He enclosed the wheel so water would flow through the waterwheel, rather than past it. This idea was developed into the water turbine, patented in 1838 by American Samuel Howd. Built on the same principle as the waterwheel, the turbine uses a higher percentage of water power. The waterwheel catches as much as it can from an open flow. The turbine focuses every drop of potential power in a closed tube. Later, James Francis curved the blades. This improved version is known as the Francis turbine. It is still in use today.

For a while, the Francis turbine was a huge part of American industry. It was briefly eclipsed by the invention of the steam engine. The birth of electricity brought it back as a highly efficient producer of hydroelectric power.

NIAGARA FALLS

In North America, Niagara Falls, between America's New York State and Canada's Ontario Province, was the birthplace of hydroelectric power. On July 24, 1880, the Grand Rapids Electric Light and Power Company used a water turbine to generate electricity. It created enough power to light sixteen brush-arc lamps. One year later, in 1881, hydropower lit the street lamps of the city of Niagara Falls.

The fantastic natural power of the Niagara River's waterfall made it an obvious location for hydropower. For more than a century, the falls had supported industrial water power. As early as 1757, there was a sawmill at Niagara. The transformation of the falls into the world's largest hydroelectric plant took both money and backbreaking labor. Hundreds of workers hollowed the bedrock of the falls to create a pit for the turbines. To divert the runoff water spilling out of the turbines, they dug a channel under the city.

An aerial view of Niagara Falls. Up to 750,000 gallons of water a second are diverted underground to the Niagara Power Project to generate hydroelectric power, one of Niagara's most important products. In addition to providing electricity, this process helps to slow the natural erosion of the falls.

In 1892, the United States and Canada formed the Canadian Niagara Power Company. Promising to protect and maintain the surrounding environment, the company built a powerhouse inside the Niagara Falls Park and diverted river water to its power plant. In exchange, the power company provided the park with the funds it needed to became a nature preserve and tourist attraction.

Located on the Ontario side of the Niagara River, the plant guides water into a holding bay and then into the brick and marble powerhouse. Once inside, it drops more than 120 feet through ten-foot-wide penstocks. At the bottom, the water rushes into a turbine.

DID YOU KNOW?

Electrical power is measured in watts. In nineteenth-century England, the ability to do work was originally described in units of horsepower. A five-horsepower motor could do the work of five horses. One horse-power equals 745.7 watts. The average table lamp uses a lightbulb of 60 watts, which means that a horse could power it using about a twelfth of its strength. Modern hydropower produces such vast amounts of energy that it is usually calculated in kilowatts (1,000 watts), megawatts (1 million watts), or gigawatts (1 billion watts).

The spinning turbine turns a three-foot-thick steel shaft, carrying power to the generators, which spin at the furious speed of 250 times a minute. As the generator creates electricity, used water spills into a discharge tunnel nearly half a mile long, which runs under the park and empties into the Niagara River below the Canadian falls.

The power company opened for business in 1905. It used two generators to produce 8,500 kilowatts of electricity, enough to power 850,000 100-watt lightbulbs. This wealth of inexpensive power created a new economy. Niagara developed into a center for electrochemical and electro-metallurgical companies. By 1924, the power company had eleven generators. In total, building the power-house cost $5,199,827.78. Today, according to the Ontario Waterpower Association (OWA), Niagara Falls produces almost one-quarter of Ontario's electric power.

Open to the Public

A non-power-related benefit of many American hydropower projects is recreation. Hydropower projects enlarge lakes or create reservoirs perfect for boating, fishing, water-skiing, and duck hunting. The land around a plant is often open to the public for hiking, skiing, and snow-mobiling. Sometimes these areas include picnic sites, swimming beaches, hiking trails, and fishing docks. Some rangers claim that hydroplants have improved the qual-ity of the fishing in the area. The Wisconsin Valley Improvement Company reports higher catch rates for smallmouth bass on hydropower reservoirs than on natural lakes.

Into the Twentieth Century

In 1882, before the Niagara plant was up and running, Wisconsin established the world's first hydroelectric station. It produced 12.5 kilowatts of power. By 1886, there were forty-five water-powered electric plants in the United States and Canada. In 1887, the town of San Bernardino, California, built the first hydroelectric plant in the West. By 1889, two hundred United States electric plants used water power for some or all of their generating needs. Less than

The Wilson Dam in Tennessee, shown here while still under construction, was an early part of the Tennessee Valley Authority Plan that tapped into the hydroelectric potential of the Mississippi River.

twenty years later, in 1907, hydropower accounted for 15 percent of electricity generated in the United States. Thirteen years after that, in 1920, the number had risen to 25 percent.

Congress passed the first federal Water Power Act in 1901. At this time, no one could build or operate a hydroelectric plant on a stream large enough for boat traffic or on federal land without a special act of Congress. During World War I (1914–1918), the nation's electrical needs increased dramatically. Congress decided to ease its restrictions. It put the secretaries of war, agriculture, and the interior in charge of the Federal Power Commission (FPC). This group could license private hydropower projects to be built on major rivers and government lands.

In the 1930s, the United States focused on hydropower development. In 1933, the Tennessee Valley Authority took charge of the hydroelectric potential of the Mississippi River in the Tennessee Valley. By 1936, there was a major hydroelectric plant at the Hoover Dam on the Colorado River. Using multiple Francis turbines, the Hoover Dam plant produced up to 130,000 kilowatts of power. In

HYDROPOWER AND TERRORISM

On October 10, 2001—almost exactly one month after the September 11, 2001, terrorist attacks on New York's World Trade Center and Washington's Pentagon—Congress met to discuss how terrorism might threaten the nation's water and energy supplies. Officials discussed government protection of dams and reservoirs. If an attack damaged or destroyed a major dam, it would have a disastrous effect on the farms and cities downstream. It could also mean the loss of hydroelectric plants and the large amounts of energy they supply to homes and businesses. Groups such as the Environmental Protection Agency (EPA) support America's counterterrorism program by helping state and local groups plan for emergencies. The Federal Bureau of Investigation (FBI) identifies weaknesses in dams and reservoirs that can make them vulnerable to attack. It also provides civilians and businesses with information about the potential for terrorism.

1937, the first federal dam, the Bonneville Dam, began operation on the Columbia River.

Engineers developed methods for storing water in order to control the supply of electricity more precisely. They used the turbines as pumps. During slow periods, the turbines pumped water to uphill storage tanks. When electrical needs rose, the water ran down into the turbines again and provided more power. The electricity produced by these dams traveled along high-voltage lines to major American cities. The success of these plants inspired similar projects in countries around the world.

By 1940, 40 percent of American electricity came from hydropower. Between 1921 and 1940, this number tripled; it almost tripled again between 1940 and 1980. In 1977, Congress abolished the FPC and shifted its duties and privileges to the brand-new Federal Energy Regulatory Commission (FERC). Nine years later, in 1986, Congress passed an act designed to control the power of FERC. Called the Electric Consumers Protection Agency (ECPA), it made FERC responsible to the recommendations of national and local fish and wildlife agencies. Today, the government officially requires FERC to value the environment (and civilian recreation) as highly as it values the potential for power when it makes power-development decisions.

3

Dams are indispensable to the world's production of hydropower. The largest hydroelectric plant in the world is the Itaipu Dam, located between Paraguay and Brazil on the Parana River. The Itaipu uses eighteen turbines to produce 12,600 megawatts of electricity, enough to light 120 million 100-watt lightbulbs all at once. Remarkable as this is, dams are usually built for reasons other than power production.

Of the 80,000 dams in the United States, fewer than 3 percent are hydroelectric. Most are used for flood control and irrigation. The United States Department of Energy (DOE) Hydropower Program lists the primary purposes and benefits of American dams as 35 percent recreation, 18 percent stock ponds, 15 percent flood control, 12 percent public

Generators inside the Glen Canyon Dam, which controls the flow of water in the Colorado River near Page, Arizona

water supply, and 11 percent irrigation. Hydroelectricity is less than one-third of the remaining 9 percent.

A dam's basic task is to control the flow of water in rivers and streams. Ancient dams halted floods and created reservoirs that provided irrigation and drinking water. In 2,900 BC, to prevent flooding, the Egyptian city of Memphis built a dam across the Nile River. The dam was a success, and the use of dams became common in the ancient world.

In the early years of the first millennium, the Romans, who were brilliant engineers, built many dams. Two thousand years later, two of these are still in use. For the Western world, the collapse of the Roman Empire meant the end of the need and the money for such major civic projects as dams. Much of the technical knowledge that had created the Roman dams was lost. It was not until the end of the nineteenth century that the need for large dams once again coincided with the ability to build them.

Today, most dams are built of concrete or earth. Earthfill dams are made out of compacted earth around clay centers. Many of these are gravity dams. The simplest type of dam, the gravity dam generally arches toward the body of a reservoir. This curve diverts the immense force of a river's water pressure back toward the shoreline. To offset the greater water pressure at the bottom of a lake or a reservoir, the concrete dam widens underwater.

A dam is either an overflow dam, in which case water spills over its rim, or a nonoverflow dam. Overflow dams have strong crests that water cannot erode. Nonoverflows use spillways to release the pressure of excess water. Spillways are either tunnels through a dam wall or hollowed channels on the crest of a dam. They provide dam operators with a tool for controlling the potential energy in the water behind a dam. Spillways also enable dam workers to divert a river. By forcing part or all of a river to flow in a new direction, engineers can provide water power to a plant located a distance away.

THE HOOVER DAM

The most famous American dam is probably the Hoover Dam. Located between Arizona and Nevada, in the Black Canyon, the Hoover Dam spans the Colorado River. The Colorado is enormous. Swollen with Rocky Mountain snowmelt, it used to thunder past the bone-dry American Southwest and empty into the Pacific Ocean. The Hoover

In addition to being a major source of hydroelectric power, the Hoover Dam, shown here under construction in 1934, contains Lake Mead, the largest reservoir in the United States. Water from Lake Mead is used to irrigate more than 650,000 acres in southern California and Arizona, and more than 400,000 acres in Mexico.

engineers wanted to divert this freshwater into the arid Colorado Basin. They succeeded. The Hoover project created a new agricultural region powered by the irrigation and electricity from the dam.

Growing crops had been nearly impossible. From 1920 to 1923, engineers sought an appropriate site for the Hoover Dam. In April 1923, they chose the Black Canyon. (They planned to call the dam the Boulder Dam, but instead it was named for Herbert Hoover, then secretary of commerce, without whose help the project would have failed.)

The Hoover is a concrete arched-gravity dam. The shape of the arch contains the enormous force of the Colorado River water in the reservoir. The dam is one-quarter of a mile long and wide enough for four lanes of traffic. The artificial lake is more than 100 miles long and, in places, nearly 600 feet deep. When it was finished, it was the largest reservoir in the world.

Construction began on July 4, 1931. The project administrators built a new town called Boulder City to house the workers and their families. Five thousand people at a time worked on the dam. First, they drilled and dynamited four tunnels through the canyon walls and around the site of the dam. This construction took a year. Once the workers had diverted the river through these tunnels, they pumped out the dam site and began to build.

By June 1933, the site was down to bedrock, and the workers poured the first concrete. Concrete takes time to cool and cure, or harden. Left on its own, the concrete dams would have taken 150 years to cool. To speed the process, the engineers invented a refrigerated-water cooling process. They surrounded the hot concrete with cooling equipment. This innovation did the job in just under two years. Once cooled, concrete continues to cure and strengthen. Even today, the concrete in the Hoover Dam grows stronger every year.

On February 1, 1935, the crew sealed the diversion tunnels. The reservoir began to fill. In the fall of 1936, three generators began to operate, one in October, one in November, and one at the end of

This photo of the Hoover Dam, shot at night, captures its magnificence. At 726 feet high and 1,244 feet long, the Hoover has been described as one of the greatest engineering works in history.

December. Eventually, fourteen more were added. The last and seventeenth began on December 1, 1961. All seventeen generators produced more than 2 million kilowatts of power. By the late seventies, the Hoover Dam had produced 150 billion kilowatts, enough power for one million people for twenty years. It would have taken 258 million barrels of oil to do the same job.

Today, the Hoover Dam still has seventeen generators with capacities ranging from 61 megawatts to 130 megawatts for a total capacity of 2,074 megawatts of power. Now the world's fifty-fourth largest dam, the Hoover Dam will always be the dam that showed the world that it was possible to create a massive hydroelectric power source.

The Three Gorges Dam

In the years since the building of the Hoover Dam, understanding of the environment has grown significantly. It was believed that the Hoover Dam would do little damage. Almost nobody was displaced from the desert zone of the Colorado Basin. At the place where the Colorado flows into the sea, however, it destroyed a precious Mexican delta jungle. Today, it is impossible to begin a dam project without raising the eyebrows and objections of commentators and environmentalists around the globe.

China's Yangtze River is 3,937 miles long, the third-longest river in the world. For centuries, the Chinese have used it as a major thoroughfare for travel and transport of goods. In 1994, work began on a

This picture shows a portion of the construction site of the Three Gorges Dam in China. As planned, the capacity of Three Gorges Dam, 17 million kilowatts, will top by roughly 40 percent that of the largest dam currently operating.

two-decade project to dam the Yangtze a thousand miles upriver from coastal Shanghai. At this spot, there are three breathtaking gorges, for which the dam will be named. When completed, at a cost of $20 billion to $100 billion, the Three Gorges Dam will be the largest hydroelectric project in the world. It will generate one-ninth of China's power.

In addition to its promise as an energy source, Chinese authorities believe the dam will control the Yangtze. One of the world's most violent and unpredictable rivers, the Yangtze killed more than one million people in the twentieth century. Its floods habitually wreak havoc on 400 million people. The dam will tame the river, allowing 10,000-ton oceangoing vessels to sail all the way into the heart of China. This will bring traffic and business to many small, upriver cities and towns.

The Three Gorges Dam is an ambitious project. It will be a mile and a half across and 600 feet high. Its lake will be 350 miles long. This is about the distance between Los Angeles and San Francisco. To create a temporary basin for the river water, some of the project's 20,000 workers are dynamiting a mountain and reusing the stone to build the dam.

Not everyone is happy about the Three Gorges Dam. The dam's 350-mile reservoir will cover 140 towns, 300 villages, and hundreds of miles of farmland. Two million people will lose their homes. The lake will submerge holy sites, ancient temples, and graveyards. It will flood a series of dramatic canyons, which have long been tourist attractions. Early in the project, the Chinese government silenced citizens who spoke out against the dam. A writer named Dai Qing, who won an award for the pieces she wrote to protest the dam, was imprisoned for ten months. Dai supports the construction of smaller, less dramatic projects on Yangtze tributaries.

In 1999, Chinese leaders began to regard the dam project more critically as they listened to the objections of people like Dai and consulted international experts. Still, China burns fifty million tons of coal a year. Any environmental damage done by the dam, say Chinese officials, will be hugely outweighed by the environmental benefits of using a clean, safe, sustainable energy source like hydropower.

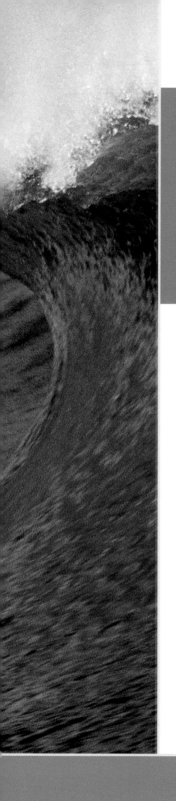

4 SEA POWER

Not every country has the natural resources for hydroelectric power. Many parts of the world are dry, flat, and riverless. Of course, there are bodies of water other than rivers. Oceans cover a little more than 70 percent of Earth's surface, and, as sailors know, there is enormous force in the crash of a wave or the swell of the sea. Tapping such a chaotic energy source is an even greater challenge than harnessing a river.

When wind creates waves in the ocean on a breezy day or during a hurricane, it transfers energy to the water. Several types of wave energy systems can capture and use this energy. The oldest and simplest sea power generators use the surge and suck of waves close to shore. They grab waves on

Small mills on Cephalonia Island in Greece derive power from unusual currents that flow through natural funnels known as swallow holes.

the way up the beach and let the sea suck them back through a waterwheel or a turbine. Newer mechanisms catch waves at their peak height and position them above large funnels. On its way down the beach, a wave rushes through the narrow neck of the funnel and powers a turbine.

It is also possible to retrieve wave energy through the use of an oscillating water column (OWC). A wave hits an OWC, compresses the air inside it, and forces the air through air turbines. Ideally, these columns are designed with two-way air turbines so that they produce power as the air forces its way in and then sucks back out of the column. Although most OWCs are still experimental, there have been some successes. In Japan, an OWC system powered a lightbulb on a navigational buoy.

One open-ocean method uses "salter ducks" to milk the power of wind-driven waves. Salter ducks are linked floats that come in chains of twenty-five. As each of the floats bobs up and down independently

5 THE GEOGRAPHY OF HYDROPOWER

Today, the world's top ten generators of hydroelectric power are Canada, the United States, Brazil, China, Russia, Norway, Japan, India, France, and Venezuela. These countries all have hilly or mountainous regions with large rivers and reliable rainfall. Successful hydropower demands appropriate geography. From a cost standpoint, a problem with hydropower is the need to own the right land.

Asia is currently home to the world's fastest-growing hydroelectric industry. In 1997, all of Asia had a total hydroelectric capacity of only about 100 gigawatts. By 2010, if projects in Japan, China, Korea, India, Myanmar, the Philippines, Indonesia, Thailand, and Vietnam reach completion, the total could be close to 300 gigawatts.

THE WORLD'S TOP TEN GENERATORS OF HYDROELECTRIC POWER

Canada: 350 billion kilowatt hours (approximately)
United States: 345 (approximately)
Brazil: 260 (approximately)
China: 180 (approximately)
Russia: 155 (approximately)
Norway: 105 (approximately)
Japan: 75 (approximately)
India: 70 (approximately)
France: 65 (approximately)
Venezuela: 60 (approximately)

China has the richest water-power resources in the world. The Chinese are currently working on the Yangtze River's 18.2 gigawatt Three Gorges Dam, the 3.3 gigawatt Ertan, and the 1.8 gigawatt Xiaolangdi projects. Altogether, this will double the country's current hydropower capacity. In contrast, Japan has already maximized its river resources and is turning to the sea. Japanese ocean-powered plants pump seawater up onto coastal cliffs to create the water drop necessary for hydropower.

The number of good hydropower sites limits a country's ability to expand hydropower production. The best American sites—on the

The Grand Coulee Dam, in Washington State, is one of the largest power dams in the United States.

Colorado, the Columbia, and the Mississippi Rivers, and at Niagara Falls—are already in use. In America's Pacific Northwest, hydropower from the Columbia River provides a more efficient power source than fossil fuels. There are 80,000 dams in the United States. Currently, only 2,400 of them generate electrical power. The rest are used for irrigation, flood control, or recreation. Power companies could easily outfit these dams with hydroelectric generators that could produce small amounts of power for inexpensive local use.

Globally, hydropower is used to generate about 20 percent of all electricity. In the United States, it produces 10 percent. In the rainy states of Oregon and Washington, it produces 85 percent. According to the United States Department of Energy, while 10 percent of energy comes from hydropower, 55 percent comes from coal, 22 percent from nuclear power, 10 percent from natural gas, 2 percent from petroleum, less than 1 percent from geothermal power, and less than 1 percent from solar, wind, and biomass power combined.

Today, the United States is capable of generating about 75 million kilowatts of hydroelectric power. This equals the output of seventy large nuclear power plants.

SMALL-SCALE HYDROPOWER

Although hydropower does not pollute the air, it can harm the environment in other ways. Large-scale projects flood farmland, displace residents, and alter the habitats of plants and animals. Both environmental and human rights groups are fighting to stop China's Three Gorges Dam project. In contrast, smaller projects offer the benefits of hydropower with fewer drawbacks. Small hydropower projects can be micro (less than 100 kilowatts), mini (100 kilowatts to 1 megawatt), or small (1 to 10 megawatts).

Micro systems are usually "run of the river" systems, which do not require the construction of a dam or a diversion channel. They use river water as it flows in its natural channel. This has almost no effect on the environment. It does not affect seasonal changes in river flow or cause flooding. Small systems rarely demand expensive equipment because they do not need to fight natural water patterns.

Chinese premier Li Peng meets with a group of villagers who will be relocated to make way for the Three Gorges Dam on the Yangtze River. By the time the project is completed, more than 2 million people will have been relocated.

Less equipment means less wear and breakage as well as less expense and more reliability. But without a reservoir, there is no way to save power. When the river slows or stops in the dry season, the system cannot produce electricity.

Micro systems are most common in small, rural communities. Farmers with water sources sometimes create their own hydroelectric systems. This provides them with low-cost, personal power. In the United States, many of these systems exist independently of government-regulated power companies. They are "off the grid," which makes them hard to document. According to a mid-1990s reckoning by the Australian Greenhouse Office, the installation costs of these off-the-grid plants come to roughly $1,500 (American) per kilowatt.

In the United States, dropping off the grid is voluntary. In some countries, large regions have no access to the electrical grid. In 1998, the Indonesian government began a project to provide 18,600 villages with power using small hydropower systems. In Vietnam, 3,000 small systems will create power for two million households. Hilly, river-laced Nepal has an installed micro hydropower capacity of about 8.7 megawatts. Officials estimate its total potential to be 42 megawatts. China currently raises its small hydro capacity by one gigawatt a year in rural areas.

Much of the future of hydropower may be in small-scale projects. The modern American community demands immense energy sources, but for daily life, most people need and use small amounts of power. The production of power as needed, by those who are using it, is both efficient and sensible. The future may see an increasing divide between the large scale of industrial power and the small scale of householders running their homes on such renewable resources as solar power, wind energy, and hydropower.

THE HYDROELECTRIC SHOE

Small hydropower systems do not have to be tiny versions of large hydropower dams and diversion projects. They can take completely different forms. The large volume of water necessary for major hydropower projects is cumbersome and restrictive. Engineers can be much more inventive using smaller amounts of water.

In Ontario, Canada, an inventor named Robert Komarechka has created a hydroelectric shoe. He noticed the smooth, repetitive, wavelike motion of the human walk. People land on their heels and roll up onto their toes. This means that the energy of the step transfers from heel to toe to heel. It occurred to Komarechka that walking on a liquid sole might provide enough energy to power a portable device like a personal stereo.

Following basic hydropower principles, he designed a shoe sole that uses tiny turbines to generate hydropower. Inside the sole, two small plastic pouches hold electrically conductive fluid. One is located under the heel and one fits under the ball of the foot. Conduits connect these pouches to a tiny generator that sits between them. A rotor with miniscule vanes drives a shaft that turns the generator. A socket on the outside of the shoe provides an outlet for plugs. As the wearer walks, fluid pushes back and forth along the bottom of the foot. A steady walk produces a steady flow of electricity. Users can access the current either through a socket on the shoe or through a device that attaches to a belt.

Ideally, hydroelectric shoes will generate enough energy for mobile phones, GPS receivers, and portable computers. In May 2001, Komarechka patented his invention as "footwear with hydroelectric generator assembly."

Hydropower can be a cheap and efficient way to produce electricity today. Maintenance costs are low. There are few pollution dangers. Once a plant is up and running, the water flow that powers it is free. Although the hydropower process does not consume the water, it is an inescapable truth that nothing is free.

Hydro development is growing more expensive. New hydropower can be more expensive to build than ever before because developers have already used most of the obvious locations. Increasingly, new sites are either dangerous or difficult to reach. Rivers that drop down sheer rock faces may provide tremendous head, but they cost

Global Warming

Hydropower does not contribute to global warming, but it will suffer from it. Less snowmelt runoff means lower reservoir levels and less water pressure in hydroelectric systems. A 10 percent runoff drop in the Colorado River means a hydroelectric power production drop of 36 percent.

more money (and lives) to develop than do waterfalls with wide, grassy banks.

Despite high initial costs, governments often opt for large-scale projects with massive dams and sizable reservoirs. They assume that, in the long run, large projects will provide more power for less money. This is not always true. Very large dams can create more expenses. For instance, many major projects necessitate canals for water distribution. They also create larger reservoirs, which demand the resettlement of more people.

Finally, dams do not last forever. Rain forest dams are generally shorter-lived than other dams. The average life of a rain forest dam is less than fifty years. Rain forest reservoirs suffer from siltation, or the accumulation of dirt. Eventually, they turn into swamps and cannot provide the flow necessary for hydropower. When a dam costs more to run than to abandon, its owners must dismantle it.

Reckoning numerical profits and costs is easy. Determining social goods and ills is harder. Hydro development often takes people's land

and destroys their lifestyles. In the Tennessee Valley, the construction of artificial lakes displaced thousands of families, flooded farms, and turned towns into "lost Atlantises."

Hydropower plants also cause environmental problems when they affect water temperatures, mineral deposits, and water aeration. Damming rivers disrupts wildlife habitats and prevents fish from swimming upstream. Sometimes these problems can be solved through dredging and the use of fish ladders, which help fish jump over dams to complete their life cycles. Other times, such measures are not enough.

Unlike many dams, the Hoover Dam did little damage to people. In the arid countryside surrounding the Colorado River Basin, there were few farmers to displace. Instead, an invisible tragedy occurred beyond the American border. The Hoover Dam interrupted the ancient progress of the silty Colorado to a Mexican delta at the rim of the Pacific. Historically, the Colorado had collected thousands of miles of mineral-rich soil on its journey across America. It dumped all of this into a lush mangrove jungle in Mexico. Before the Hoover Dam, this forest was home to birds, ducks, fish, quail, bobcats, and coyotes. When the river mud dried up, the area became a desert. This kind of ecological destruction is not unusual.

Hydro development causes major deforestation in tropical rain forests. On the island of Borneo, one-quarter of a precious chunk of the world's remaining rain forest is threatened

Sacred land of native peoples, such as this burial ground in Malaysia, is often threatened by ambitious hydro projects.

by hydropower projects. According to the United Nations Food and Agriculture Organization, dams on tropical rivers cause 10 percent of the total annual deforestation in the Tropics. In some areas the numbers are higher. Part of the problem is caused by support projects. Building a dam means building roads, cutting paths for power lines, using heavy construction machinery, and forcing human migrations. Workers move to work sites, and local people move away from their homes.

The people who live on the land before a project begins are not generally consulted or paid for their land. They are either "resettled" or forcibly moved. This movement creates new housing and work needs. It also lowers local food production. Workers who staff the projects often have different geographical backgrounds and immune systems. Sometimes they introduce diseases that are locally debilitating or deadly.

A Healthy Environment

Some people believe that human needs are more important than the health of the environment. Biologists Paul and Anne Erlich have a response for these people. They suggest that destroying species may be the same as slowly pulling the bolts from a plane wing. At first, nothing happens. Suddenly, a wing snaps off, and the plane hits the ground. In their view, it is hard to know how many species the planetary ecosystem may need to stay aloft.

THE ECOLOGICAL BALANCE

Interruption of the cycle in the rain forest tends to kill plant and animal species. Scientists believe there are between five and one hundred million species on the planet. Of these, they have identified only one and a half million. Between 50 and 90 percent of these species probably live in tropical rain forests. The undisturbed rain forest is a very stable, very fertile environment. It is wet. The temperature is constant. For sixty million years, there was little human activity. Now that has changed.

In the human view, a hydropower plant is a clean, inexpensive, long-term solution to growing energy needs. For a sixty-million-year-old rain forest, a hydropower reservoir that may last fifty years is not a long-term solution. From the ecological perspective, it may be poor judgment to privilege human claims over those of a fragile, diverse environment that affects all life on Earth.

Even from a human-centric, pro-hydropower perspective, the death of the rain forests will have a mammoth impact on the global environment. The hydrological cycle depends on balances of heat and rain. The water that falls in the rain forest evaporates and returns to the atmosphere almost immediately. The rain forests help to keep the world wet. As rain forests vanish, regional climates grow dryer. When this happens, hydropower facilities become less productive.

Hydropower is a huge global business. The World Bank estimates that about $100 billion was spent building hydropower projects at the end of the twentieth century. While many people see the benefits of hydropower, now is the time to recognize the drawbacks.

To use hydropower without safeguarding the environment is not so different from using fossil fuels. In theory, water power may be sustainable, but in the long run, water is not. It suffers from development, pollution, and global warming. Hydropower is only sustainable if we maintain the environment that makes it possible. The challenge for the future of hydropower is to learn to treat the water and the world as a precious interconnected whole.

GLOSSARY

alternative energy Energy derived from sources that do not use up natural resources or harm the environment.

biodiversity The number and variety of organisms within a geographical area.

biomass Plant material or agricultural waste used as a fuel or energy source.

dam A barrier constructed across a waterway to control the flow or raise the level of water.

deforestation The removal of trees.

ecology The science of relationships between organisms and their environment.

ecosystem The ecological community and environment functioning together.

estuary An arm of the sea that extends inland to meet the mouth of a river.

flow The amount and speed of water entering a water-wheel or turbine.

fossil fuel A hydrocarbon deposit such as coal or oil derived from living matter of a previous geologic time.

generator A machine that converts mechanical energy to electrical energy.

geothermal Of or relating to the internal heat of Earth.

head The height or distance between the high point of a body of water and the point at which energy is drawn from it.

hydroelectric Generating electricity by conversion of the energy of running water.

hydropower The use of water to create power, generally hydroelectric power.

kilowatt A unit of power equal to 1,000 watts.

penstock A pipe used to carry water to a waterwheel or turbine.

reservoir A natural or artificial pond or lake for storing and regulating water.

runoff Rainfall or snowmelt that is not absorbed by the soil.

silt Mud or fine earth deposited by running or standing water.

spillway A channel for overflow of water from a reservoir.

stator A stationary part of a motor about which a motor revolves.

tidal power Hydropower derived from the rise and fall of the tides.

turbine A waterwheel with a series of curved blades or buckets that converts the kinetic energy of a moving fluid to mechanical power.

FOR MORE INFORMATION

Federal Energy Regulatory Commission (FERC)
888 First Street NE
Washington, D.C. 20426
(202) 219–2991
Web site: http://www.ferc.fed.us/hydro/docs/
   waterpwr.htm

WEB SITES

Due to the changing nature of Internet links, the Rosen
Publishing Group, Inc., has developed an online list of
Web sites related to the subject of this book. This site is
updated regularly. Please use this link to access the list:

http://www.rosenlinks.com/lfe/hypo/

FOR FURTHER READING

Arnold, Guy, and Peter Harper. *Facts on Water, Wind and Solar Power*. New York: Franklin Watts, 1990

Bailey, Donna. *Energy from Wind and Water*. Austin, TX: Raintree/Steck-Vaughn Publishers,1990.

Dunn, Andrew, and Ed Carr. *The Power of Pressure*. Stamford, CT: Thomson Learning, 1993.

Lewis, Alan. *Water*. New York: Franklin Watts,1980.

McClymont, Diane, and Richard Young, ed. *Water*. Ada, OK: Garrett Educational Corporation,1991.

Twist , Clint. *Wind and Water Power*. New York: Franklin Watts, 1993.

BIBLIOGRAPHY

Bernstein, Paula. *Alternative Energy: Facts, Statistics, and Issues.* Phoenix, AZ: Oryx Press, 2001.

Echeverria, John D., Pope Barrow, and Richard Roos-Collins. *Rivers at Risk: The Concerned Citizen's Guide to Hydropower.* Washington, DC: Island Press, 1990.

Eshenaur, Walter. *Understanding Hydropower.* Scottsdale, AZ: Vita Publications, 1985.

Hazen, Mark E. *Alternative Energy: An Introduction to Alternative and Renewable Energy Sources.* Indianapolis, IN: Prompt Publications, 1996.

Potts, Michael, and John Schaeffer. *The Independent Home: Living Well with Power from the Sun, Wind, and Water.* White River Junction, VT: Chelsea Green, 1993.

Watts, Martin. *Water and Wind Power.* Buckinghamshire, United Kingdom: Shire Publications, 2000.

INDEX

CREDITS

ABOUT THE AUTHOR

Allison Stark Draper has written books for young readers about science, history, and the environment. She lives in New York City and in the Catskills.

PHOTO CREDITS

Cover © Michael T. Sedam/Corbis; p. 4 © Gary Braasch/Corbis; p. 6 © Jeffrey Aaronson/Timepix; pp. 7, 10 © Corbis; p. 8 © Dean Conger/Corbis; p. 12 © Bob Rowan/Progressive Image/Corbis; p. 14 © Ron Watts/Corbis; p. 16 © Schenectady Museum/Hall of Electrical History/Corbis; pp. 18, 28, 32, 47 © AP/Wide World Photos; p. 20 © Rick Doyle/Corbis; p. 21 © Hulton/Archive by Getty Images; p. 24 © Eye Ubiquitous/Corbis; p. 26 © Ric Ergenbright/Corbis; p. 30 © Charles O'Rear/Corbis; p. 34 © David Pu'u/Corbis; p. 36 © Gail Mooney/Corbis; p. 37 © Morton Beebe/ S. F./Corbis; p. 42 © Galen Rowell/Corbis; p. 45 © Charles E. Rotkin/Corbis; p. 50 © Robert Nickelsberg/Timepix; p. 54 © Robin Moyer/Timepix.

LAYOUT AND DESIGN

Thomas Forget